The Root Manifesto

Reclaiming Cybersecurity in a Compromised World

Christopher Quinn

PURPLE TEAM SECURITY

The Root Manifesto: Reclaiming Cybersecurity in a Compromised World

ISBN: 979-8-9988306-7-9

Publication Date: July 4, 2025
Published by **Purple Team Security**

10 9 8 7 6 5 4 3 2 1

Impressum

Autor: Christopher Quinn
Verlag: Purple Team Security
Adresse: 6339 Charlotte Pike, Unit #B382, Nashville, TN 37209
E-Mail: cquinn@purpleteamsec.com
Verantwortlich für den Inhalt nach § 5 TMG: Christopher Quinn
ISBN: 979-8-9988306-7-9

Preface

This is not a book about
compliance.

It is not a checklist, a framework,
or a set of corporate platitudes
pretending to be strategy.

This is a declaration. A gauntlet.
A line in the sand. A refusal—written
in root shells and rage.

Currently, trust is extracted,
privacy is bled, and security has
been transformed into a service-

level illusion. Where billion-dollar breaches are met with billion-dollar shrugs. Where C-suites speak of "transparency" while redacting every meaningful log. Where defenders scream into silenced SIEM dashboards while vendors peddle hope with a login page and a liability waiver.

This is a time where caring is punished. Where questions are quarantined. Where ethics are labeled as "non-strategic." Where compliance becomes the floor and the ceiling. And truth—raw, urgent, inconvenient truth—is buried beneath dashboards, diluted by metrics, and drowned in the politics of posture.

Enough.

This manifesto is for the ones they call paranoid. For the sysadmin rebuilding the same broken system every month because no one will fund the fix. For the red teamer who sees just how brittle our bastions are. For the SOC analyst whose fatigue is weaponized as a "workforce gap." For the incident responder whose tools are throttled, whose alerts are ignored, and whose burnout is rebranded as "attrition."

It's for the ones who encrypt not out of secrecy—but sovereignty. Who run Tails because they know better. Who write their own scripts, host their own infra, and log everything not because they mistrust everyone—but because they understand systems decay.

It is for the weary. The wary. The watchful. For those who refuse to normalize neglect. For those who would rather be inconvenient than complicit. For those who remember that behind every "acceptable risk" is a human being who was never consulted.

This book does not seek your permission.

It seeks your alignment.
Your rage.
Your refusal.

It offers no silver bullets.
No vendor logins. No "modern XDR-powered zero-trust orchestration platforms."

It offers something more dangerous:
Perspective.
Principle.
Provocation.

Defend what matters. Reject what doesn't. Own what you build. Understand what you use. Pass the torch before it's extinguished.

You will not agree with every word. Good. That means you're thinking.

This is not a neutral book. It is not gentle. But it is honest.

And in security, honesty is in short supply.

So read this not as doctrine—but as fuel. Not as gospel—but as grit. Not to be admired—but to be acted upon.

This is not just a preface. It's a permission slip to burn the old playbook. To rebuild with principle. To resist the rot.

We are the firewall now.

Let's act like it.

And if they tell you to tone it down?

Smile.

Then turn the volume up.

Contents

Chapter 1

A System That No Longer Deserves Our Trust

"You are either root, or you are being rooted."

— Anonymous

We are told to trust the system.

But the system outsourced its brain, sold off its soul, and now demands that we pretend it still deserves our

faith. Institutions built to protect
data have become its biggest threat.
Agencies assigned to safeguard our
liberties have become entangled
in the surveillance business. The
platforms we depend on now depend on
us remaining dependent. This is not
paranoia. This is patch notes and
breach disclosures, leaked memos and
whistleblower accounts. It is reality,
and it is worsening.

Since 2013, we've been told Snowden
was a turning point. A wake-up call.
But a decade later, the surveillance
has only become smoother, cheaper,
and more marketable. It's not just
governments anymore—it's gig apps,
SaaS dashboards, even school-issued
Chromebooks. Every convenience bleeds
telemetry. Every interaction is a
trade. And the ledger is never in
your favor.

We live in an age where digital
convenience is sold like sugar while
surveillance hides in the ingredients
label. Your smart TV watches you.
Your browser fingerprint lasts longer

than your mortgage. Your phone tracks more than your location—it maps your behavior, your bias, your biology. All with your passive consent. Opt-in by design, opt-out by fantasy.

They say "if you have nothing to hide, you have nothing to fear"—but they never say that to the ones doing the watching. They don't say it to the corporations harvesting your data, the agencies skimming your messages, the bureaucrats deciding what counts as a "threat."

They don't say it to the ones logging your packets, scraping your keystrokes, or siphoning your contact lists into classified databases. They don't say it to the people who invented facial recognition but won't disclose its error rates. They don't say it to the boardrooms where your privacy is monetized one demographic segment at a time.

This manifesto isn't about compliance. It isn't about best practices. It's about control—and who has it.

We build out firewalls, but forget the fire burns inside too. We encrypt at rest and in transit, but leave endpoints wide open. We trust certificates, corporations, and cloud contracts without reading the clauses that own us. We implement MFA but forget why the "A" matters—because access is power, and access unchecked becomes violation.

Security vendors want to sell you salvation. Regulators want to mandate obedience. But the real power—the kind that makes systems robust, the kind that bends outcomes—is still local. Still technical. Still yours to take, if you have the will and the skill.

This book is not a toolkit.

It's a reckoning.

Because the truth is—this system was never built for your protection. It was built for control. You, reader, are either root—or you are running someone else's process with privileges they gave you. And like any bad process, it doesn't ask. It assumes.

It assumes your consent. It assumes your submission. It assumes your silence.

And that's where it's wrong.

This is the hard truth of digital autonomy: unless you claim control, someone else will.

And let us be clear: no policy, no product, no audit report will save you when your agency chooses convenience over courage. When the people who should've protected you were too busy attending meetings about roadmaps they'll never follow. When your leadership calls silence a "communication strategy" while attackers pivot across your network.

We don't need more policies. We need purpose. We need a generation of defenders who don't just implement controls, but understand why those controls exist. Who see beyond dashboards. Who know the kernel, not just the GUI. Who fight for privacy not because it's popular, but because it's personal.

And when you fight, you will be

mocked. Marginalized. Dismissed as "paranoid," "too intense," "not aligned with culture." That's fine. Better to be called paranoid and right than passive and breached. Better to be a thorn in their side than a rubber stamp on their failure.

Better to walk alone in truth than march in unison toward breach.

This is not fear-mongering. This is realism—tempered by resolve. Because nothing is inevitable. Not the breach. Not the burnout. Not the betrayal.

So we begin here.

Not with policy. Not with passwords. Not with products.

With purpose.

The kind that defends even when no one is watching. The kind that walks when staying would mean silence. The kind that rewrites the system—not just patches it.

Because the world doesn't need more frameworks. It needs a firewall with a spine.

Chapter 2

Root or Be Rooted

"Give me control of the shell,
and I care not who writes the
policy."

— A sysadmin, probably

The most dangerous myth in modern
computing is that convenience and
control can coexist indefinitely.
That you can delegate root access to
someone else—an operating system, a
vendor, a cloud provider—and still
call yourself secure. That illusion
ends here.

To root is to reclaim.

Root is not just a user account.
It is a mindset. A commitment. A
declaration that you are no longer
willing to outsource your autonomy to
a black box maintained by someone with
different priorities, incentives, or
values. It is not for everyone. But
if you want to control your machine—
and by extension, your data, your
identity, your future—you must start
at root.

Root is accountability made
manifest. It means you cannot blame
the vendor. It means the patch
doesn't come from a portal—it comes
from you. It means when the alert
hits, there is no escalation path.
There is only you and the system you
swore to defend.

The Problem With Managed Everything

We've outsourced so much, we've
forgotten how to log in. Cloud
services abstracted so far up the

stack that most engineers couldn't survive a flat network with a bare metal box. We deploy YAML and Terraform and Dockerfiles like totems to a machine god we don't understand. And then we wonder why the breach was undetected for nine months.

Managed EDR, managed DNS, managed identities. It's like asking someone to lock your door, guard your family, but never tell you who holds the key. Ask your security team who manages your SSO metadata or your cloud KMS keys. If the answer starts with "a vendor," you are no longer secure—you are merely serviced.

Root is not about distrust. It's about verification.

Security doesn't come from the illusion of control—it comes from understanding the blast radius when that illusion fails.

Convenience is the anesthesia of compromise. It numbs us just long enough to get owned.

Reclaiming the Stack

Start with your own machine. Can you
audit your OS? Can you boot without
calling home? Can you verify your
BIOS, your kernel, your init system?
If not, you don't own it. It owns
you.

Install Linux—not because it's
free, but because it's inspectable.
Harden it—not just with scripts, but
with understanding. Ditch the bloat.
Embrace the terminal. Learn what each
daemon does, which ports open when,
and why.

Reclaim your services. Run your own
DNS. Your own VPN. Your own identity
provider, if you dare. Not because
it's trendy, but because it's yours.
Every layer you own is one less layer
someone else can exploit.

This is not about being anti-cloud.
It's about knowing when the cloud
becomes a crutch. When abstraction
replaces skill. When the ease of
provisioning becomes the root cause of
compromise.

Ask yourself—if the network died today, how many of your skills would survive offline?

Know your stack like you know your tools. Know it like you know your rifle. Because when the breach hits, you will fall back on what you truly understand—and nothing more.

Operational Sovereignty

Root is responsibility. It's waking up at 2 a.m. to patch a vuln. It's reading CVEs over coffee. It's the price of sovereignty—and it's worth it.

Because once you reclaim root, you start to see the shell for what it is: the last place they haven't monetized. The CLI is not just efficient. It is sacred. It is the narrow gate to understanding, to autonomy, to defense.

The shell doesn't lie. It doesn't auto-correct your assumptions. It forces precision. And in precision, there is power.

The moment you understand how to bootstrap a machine from bare metal to hardened host without third-party babysitting, you will never see infosec the same way again. You stop being a product. You become a defender.

To be root is to be awake.

Root as Philosophy

Root is not a badge. It's a burden.

This chapter is not a tutorial. It's an invocation.

To root is to take ownership—not just of systems, but of your role in the digital ecosystem. Of your power to secure, to resist, to build.

Root is where excuses die.

It is where the script kiddie becomes the operator. Where the corporate drone becomes the insurgent. Where the consumer becomes the creator.

Root doesn't ask permission. It

executes.

This isn't about arrogance. It's about agency. About choosing clarity over comfort. About being the kind of engineer who can't be gaslit by a GUI.

Own the system. Or be owned by it.

That's the choice. And in this book, we choose root.

Chapter 3

Compliance Theater and the Myth of Security

> "Compliance is a checklist.
> Security is a commitment."
>
> — Anonymous Auditor

Security in today's institutions has been boiled down to passing audits. Not protecting people. Not defending data. Not building

resilient systems. Just checking
boxes.

We are drowning in frameworks:
HIPAA, PCI-DSS, NIST, ISO 27001,
HITRUST, SOC 2. Each one promising
trust. Each one weaponized by
bureaucracy to substitute effort
for outcomes. The result? Breached
hospitals that were HIPAA-compliant.
Compromised fintech firms that aced
their SOC 2. Ransomware on "certified"
endpoints.

Compliance is not security. It
never was. Nor was it meant to be. It
was about control.

The Box-Checking Industrial Complex

Every time a breach hits the
headlines, someone inevitably says:
"But they were compliant." As if
that's supposed to mean something.
What it means is this: the attackers
didn't care about your policy. They
cared about your patch cycle, your
exposed RDP port, your junior sysadmin

with a weak password.

Compliance documents what security should look like. It does not verify whether that security actually exists. Worse, it incentivizes minimalism—do just enough to avoid a fine. Nothing more.

We've seen it firsthand: VPNs with unpatched vulnerabilities getting greenlit because "the policy says it's in scope". Flat networks with privileged accounts accessible from the guest Wi-Fi because "segmentation wasn't on the compliance roadmap yet."

The compliance industry is not the enemy. But it's not your friend either. It exists to reduce liability, not risk. To appease regulators, not to protect users. Never confuse legal safety with real safety.

Bureaucracy has learned to mimic security posture, not produce it. It rewards appearance, not efficacy. It throws consultants and training decks at zero-day realities, then pats itself on the back when the breach was "within acceptable tolerances."

Security as Theater

Security theater is performing
controls that look effective but
accomplish nothing. Mandatory
password changes every 90 days.
Printed-out security awareness
posters. Threat models that only
exist in PowerPoint.

We've mistaken artifacts for
impact. The policy binder is full,
but the firewall rule set hasn't been
reviewed in years. The risk register
is pristine, but shadow IT runs half
the infrastructure.

This is the infosec equivalent of
putting a "Beware of Dog" sign on a
house with no dog.

And auditors keep checking to make
sure the sign hasn't faded.

Executives recite buzzwords.
Middle management files tickets. And
engineers are told to wait until
Q4 for a security budget, while
adversaries operate on Q1 timelines
with zero procurement delays.

I once watched a major healthcare

provider spend $60,000 on a HITRUST
consultant but reject a $5,000 upgrade
to isolate legacy medical devices from
the main network. Their logic? The
consultant satisfied a requirement.
The upgrade didn't.

Breaking the Cycle

True security doesn't start with
frameworks. It starts with threat
modeling. With understanding your
assets, adversaries, and attack paths.
Then—and only then—should you map that
reality back to a framework.

   The most secure organizations
I've worked with were not the most
compliant. They were the most curious.
They tested their assumptions. They
drilled their incident response.
They knew where their sensitive data
lived and who could access it. Their
culture treated security not as a
department, but as a habit.

   Ask any mature security team how
they sleep at night. It won't be

because their audit went well. It'll
be because they've seen their own
system break, and they fixed it before
someone else did.

You can't outsource vigilance.

You can't template responsibility.

You can't audit your way into
resilience.

And if your culture punishes
urgency, shames curiosity, and rewards
passivity, then your breach is not
a matter of if—it's a matter of
scheduling.

From Compliance to Commitment

This isn't a manifesto against
compliance. It's a demand to go
beyond it. Use the framework. But
build a culture. Document your
controls. But verify them in reality.
Pass the audit—but then ask yourself:
"Would we survive the breach?"

Security isn't what you say you do.

It's what your adversary sees when
they test it.

And let's be clear—they're testing it now.

So the next time someone asks, "Are we compliant?"

Respond with a better question:

"Are we ready?"

And if the answer is no—start fixing that today.

Because while compliance might win you a contract,

Only commitment will win you the war.

Chapter 4

Privacy Is a Prerequisite to Liberty

"Arguing that you don't care
about privacy because you have
nothing to hide is like saying
you don't care about free speech
because you have nothing to say."

— Edward Snowden

There is no such thing as liberty
without privacy.

You can't speak freely if every
word is logged. You can't think freely

if your queries are profiled. You
can't dissent if surveillance systems
pre-classify you as a threat.

Privacy is not secrecy. It is
dignity. It is the right to exist
without being reduced to data points.
It is the freedom to explore, to
communicate, to evolve—without being
pre-scored by some invisible algorithm
deciding your risk, your credit, your
guilt.

In an age where every action is
tracked, every movement analyzed, and
every click commodified, privacy is
no longer a luxury. It is a survival
tactic. It is not the fringe of the
conversation—it is the conversation.

Surveillance Capitalism and the Data Economy

The internet didn't become free. You
became the product.

Advertising is no longer about
selling ads. It's about modeling
behavior. Your phone, your search
history, your map pins, your email

metadata—all bundled, auctioned, and resold with surgical precision. The trackers are silent. The consent is fake. The implications are not.

Insurance companies want your fitness data. Employers want your sentiment scores. Law enforcement wants access to everything.

And every time you install an app or sign up for a "free" service, the price is you. Your habits. Your patterns. Your digital DNA.

There's no privacy policy strong enough to overcome an economy built on extraction.

The surveillance machine doesn't need your secrets. It needs your behaviors. It doesn't care about your name—it cares about your nature. Your psychology. Your predictability. That is what's being harvested.

And the worst part? We've normalized it. We've taught a generation that this is the price of participation. That you must surrender your digital skin to access a public square. That convenience means compliance, and resistance is somehow outdated.

That's not progress. That's conditioning.

Privacy Is a Collective Defense

When people say "I have nothing to hide," they misunderstand the stakes. Privacy isn't just for criminals or whistleblowers. It's for parents. Patients. Protesters. It's for the vulnerable.

Privacy isn't personal—it's communal. It's a herd immunity against tyranny.

When you refuse to use encrypted communication, you weaken the norm. When you dismiss privacy tools as "paranoid," you delegitimize their use.

Every person who uses Tor normalizes it. Every person who encrypts their email strengthens the expectation of confidentiality. Privacy spreads through culture.

Through example. Through solidarity.

Your personal threat model might be low—but someone else's life depends on your habits.

Think beyond yourself. Think of the teenager escaping an abusive household, the journalist under surveillance, the activist crossing a border. Your operational laziness might be their operational risk.

Silence is complicity. Insecurity is contagion.

The Tools of Liberation

Use Signal. Not because you're hiding something—but because your rights are not up for sale.

Use Tails. Because a stateless OS is what freedom looks like on a USB stick.

Use end-to-end encryption, not because you fear exposure, but because you deserve control.

Teach your family about metadata. Show your kids how to generate strong

passwords. Normalize burner phones
and anonymous browsing. Not as fringe
behavior—but as civic literacy.

In a world where privacy is
abnormal, tools become rebellion.
Make them routine instead.

Run your own services. Ditch the
tracking pixels. Block the scripts.
Educate your tribe. Make privacy not
just defensible—but desirable.

Adopt default-deny not just
for firewalls—but for permissions,
platforms, and platforms that presume
ownership over your experience.

Normalize saying no. To cookies. To
TOS. To digital trespass.

The Privacy-First Mindset

Start asking:

- Who owns this data?

- Who profits from it?

- Who can access it without my
  consent?

– What would happen if it were exposed
  tomorrow?

  Security starts with these
questions. But liberty depends on
their answers.
  Privacy is not dead.
  It is being killed deliberately,
quietly, systematically.
  And every byte we reclaim is a byte
of freedom preserved.
  We fight for privacy not because
we're hiding.
  We fight for privacy because we are
free.
  Because freedom, once lost, rarely
returns without a fight.
  So fight.
  Fight in your browser settings.
In your app choices. In your
infrastructure.
  Fight by refusing to normalize
digital submission.
  Because if liberty is to survive
the algorithm, it must be encrypted
first.

Chapter 5

Own Nothing, Be Nothing

"You will own nothing, and you
will be hacked."

— A realist's response to The
Great Reset

We've traded ownership for access.
Servers for subscriptions. Knowledge
for convenience. And in doing so,
we've built a world where the average
user no longer controls anything—not
their data, not their devices, not

even their decisions.

This isn't an accident. It's a business model.

The less you own, the more you depend. The more you depend, the easier you are to monetize, to influence, to compromise. Ownership is not just a technical question—it is a political one. And in this chapter, we take it back.

The Death of Locality

There was a time when your computer was yours. Your files lived on a drive. Your software came in a box. Your choices didn't need approval from an app store.

Then came the cloud.

Now, your documents live on someone else's server. Your photos get scanned for "safety." Your productivity suite updates without asking. And your data is subject to the laws of whatever jurisdiction your provider operates in this fiscal quarter.

   You pay for access. But you don't
own a single thing.

   And when access is revoked, so
is your autonomy. Control lost by
convenience is almost never regained.

From Infrastructure to Insecurity

Every SaaS product promises speed.
Every managed platform promises scale.
But none of them promise sovereignty.

   Use a cloud IDE? You've outsourced
your compiler. Rely on third-party
APIs? You've externalized your risk.
Your infrastructure-as-code is code
you don't control—it's an abstraction
built atop a service you rent.

   And when the breach hits, the
vendor shrugs. Because their SLA never
promised security. Only uptime.

   This isn't resilience. It's
roulette.

   And too many orgs are spinning the
wheel daily, then acting surprised
when the chamber isn't empty.

Digital Tenant Farming

In this model, you don't own the land.
You rent the platform. You build
your livelihood on someone else's
soil. And when the terms change, or
the landlord sells, or the service
sunsets—you lose it all.

You are not a customer. You are a
tenant.

This isn't just bad security. It's
digital feudalism.

And the new aristocracy wears
hoodies instead of crowns.

They call it progress. But it's
dependency by design. And if you ever
challenge it, you're labeled "anti-
tech," "anti-cloud," "paranoid."

No. We're not paranoid. We're
awake. We've read the TOS. We've
watched the breaches. We've seen what
happens when a dependency chain snaps
and the whole stack crumbles like
rotted drywall.

Dependency is not a sin. But it
becomes one when you pretend it's
strength.

Reclaiming Digital Property

Start local. Run your own services.
Host your own files. Own your domain,
your email, your storage.

Buy hardware you can audit. Flash
your firmware. Kill the telemetry.
Understand what your devices are doing
when you're not watching.

You don't need to self-host
everything. But you should know how
to.

Understand the stack beneath your
software. Know what runs your server.
Know who controls your DNS.

Teach your team to build minimal,
resilient systems. Teach your kids
what filesystems are. Remind your
friends what offline feels like.

Ownership isn't extremism. It's
realism.

Because the moment the network goes
down, the cloud goes silent. And what
you have left—that's your true stack.
That's your digital spine. That's
the line between sovereignty and
subjugation.

The Cost of Convenience

Every convenience has a cost. Every abstraction hides complexity. Every dependency is a doorway.

Ask yourself: if the service shut down tomorrow, would your digital life survive?

Security without ownership is an illusion.

You cannot defend what you do not control. You cannot control what you do not own.

This Is the Line

This isn't just a call to self-host. It's a call to self-govern.

If your tools can be revoked, your voice can be silenced. If your access can be throttled, your rights can be throttled.

We will not build our futures on leased land. We will not pretend that subscription equals sovereignty. We will not trade autonomy for uptime.

So own it. Or be owned.
Because the future doesn't belong
to the most connected. It belongs
to the most sovereign. And those who
refuse to rent their resistance.

Chapter 6

Minimalism as Defense

> "Every line of code is a
> potential liability."
>
> — Every paranoid engineer, ever

In security, every component is a
potential compromise. Every plugin is
a possible payload. Every dependency
is an unvetted guest.

The more you run, the more you
expose. The more you expose, the more
you risk. It's not theoretical—it's
operational. And yet the industry
continues to build brittle castles

atop complex sand.

   We call this progress. It's
not. It's entropy masquerading as
innovation.

Complexity Kills

Attackers don't need to break your
whole system. Just one part of it.
One misconfigured reverse proxy. One
forgotten debug port. One outdated
library with an unpatched CVE.

   Modern infrastructure is
a Frankenstein's monster of
microservices, frameworks, and
containers glued together with YAML
and hope. There are too many moving
parts. Too many assumptions. Too few
defenders who understand the full
picture.

   When your architecture looks like
a Jackson Pollock painting, don't
be surprised when attackers find the
gaps.

   Security isn't magic. It's math.
And complexity is exponential—one

function away from collapse.

Every new feature is a new failure mode. Every third-party integration is a coin flip. And every abstraction layer is one more blind spot in a crisis.

The Minimalist Advantage

Minimalism is not about asceticism. It's about control.

Use fewer packages. Fewer layers. Fewer protocols. Prefer static binaries. Prefer known-good configurations. Prefer architectures you can actually audit.

A minimalist system is observable. It's explainable. It fails predictably.

It's not cool. It's not flashy. It works.

And in a crisis, that's all that matters.

Because when everything breaks—and it will—it won't be your dashboards or your vendor SLAs that save you. It

will be your understanding. Your logs.
Your shell. Minimalist systems bring
clarity. And clarity is a weapon.

Security by Subtraction

Start asking: what can we remove?

Can you replace a dynamic web
app with static HTML? Can you strip
the init system to its essentials?
Can you deploy without a full-blown
container orchestration nightmare?

Use Alpine. Use OpenBSD. Use TUI
tools. Use software written by people
who hate bloat and love control.

Your system should do only what you
tell it to do—and nothing more.

Complexity is not a sign of
intelligence. It's a sign of lazy
architecture.

Subtraction is not regression. It's
refinement.

Minimalism in Practice

Kill unnecessary services. Audit your crontabs. Verify every socket listening on your box. Run 'netstat' and ask yourself: why is that port open?

This is the practice of defensive minimalism—not just reducing surface area, but understanding what should be there, and why.

Conduct architecture reviews like war games. Assume compromise. Ask where they'd hide. Then rip it out.

Minimalism requires discomfort. It forces you to make decisions. But decisions create awareness. And awareness creates security.

Because if you're not pruning, you're rotting.

The Antidote to Overengineering

We've been sold complexity as a solution. Kubernetes for everything. Abstractions on abstractions. But security thrives on simplicity.

The fewer moving parts, the fewer failure modes. The simpler the stack, the faster the incident response. The leaner the system, the harder it is to hide an implant.

Minimalism is not fashionable. It's foundational.

Build small. Build lean. Build with the assumption that you will be attacked—and design accordingly.

Because the adversary doesn't care how elegant your stack is. They care how easy it is to break.

They don't read your README. They read your sockets.

This Is Not Optional

Every line of code is a door. Every dependency is a negotiation. Every service you forget is a beacon for someone else's payload.

Your bloat is their opportunity.

Minimalism isn't about saying no to modernity. It's about saying yes to sovereignty. Yes to clarity. Yes to the kind of control that doesn't need an API key to authorize your own defense.

Strip it down. Tear it out. Get to the bones of your infrastructure.

And then harden the bone.

Because the only thing worse than being breached is being breached by a vulnerability in a feature you never needed, didn't ask for, and couldn't disable.

Minimalism is the refusal to be owned by your own stack.

And in an era where everything calls home, calls out, and calls the shots—you had better know exactly what's running, and why.

Because in this world, ignorance isn't bliss. It's breach.

Chapter 7

Digital Autonomy Through Linux

> "Linux is not just an operating system. It is a declaration of independence."

— A sysadmin with root

To defend a system, you must first understand it. Not just what it does—but how it works, why it works, and what happens when it breaks. Linux is not merely a platform. It is a path to understanding, a gateway to control, a canvas for security.

The command line is not retro. It's not niche. It's the front door to digital autonomy.

Why Linux?

Because you can see the source. Because you can recompile the kernel. Because the init system is not a mystery but a mechanism. Because the filesystem hierarchy is legible, and the logs are yours.

Linux doesn't hide. It teaches.

And in a world of closed systems and managed platforms, that's revolutionary.

The Culture of Mastery

Linux forces you to learn. That's not a bug. It's the feature.

You learn what a process really is. What a socket really does. How memory is allocated. How permissions are structured. And once you learn these

things, you don't just become a better
user—you become a builder. A breaker.
A defender.

This is not about gatekeeping. It's
about empowerment.

Linux is hard because control is
hard. But it gets easier the deeper
you go.

Distros as Philosophy

Every Linux distribution carries a
worldview:

- **Arch**: Freedom through friction.

- **Debian**: Stability through community.

- **Alpine**: Minimalism through
  intentionality.

- **Qubes**: Compartmentalization as
  defense.

Choose the one that matches your
values. Switch when your values
evolve.

Use your distro like a toolkit, not a tribe.

Break It to Know It

The best way to learn Linux is to break it. Delete your desktop environment. Kill your network stack. Boot into single user mode. Watch what fails. Learn how to fix it.

Because when a real incident hits—when the system won't boot, when the logs are gone, when the services are crashing—you won't panic. You'll pivot.

Linux is not just robust. It's resilient. And it teaches you to be the same.

If you've never typed 'fsck' while sweating, you haven't lived.

Security Through Literacy

Security tools come and go. SIEMs get replaced. AV engines get swapped. But your ability to drop into a shell, debug a misbehaving system, and identify malicious behavior—that endures.

Learn Bash. Learn systemd. Learn iptables and nftables. Learn strace, lsof, tcpdump.

These aren't tools. They're weapons.

Linux literacy is the single greatest investment you can make in your security future.

Command Line = Control Line

GUI is what vendors show you. CLI is what the system really is.

And the moment you understand that, the moment you realize that owning your system means mastering your system, you will never go back.

Autonomy is not a checkbox. It's a terminal prompt.

And it begins with: `sudo -i`

Chapter 8

Educate or Extinguish

> "If you want to control the
> future, control how people think
> about the present."

— Unknown

We are not losing because we lack
tools. We are losing because we are
failing to teach. Failing to arm the
next generation with the intellectual
ammunition required to survive a world
where information is weaponized and
ignorance is profitable.

Education in cybersecurity has

become a parody. Flashcards over fundamentals. Certifications over comprehension. Career paths paved by LinkedIn influencers instead of lived experience. We don't need more CISSPs. We need more critical thinkers. We need defenders with grit, with scars, with a reason to fight.

This chapter is not an appeal. It is an ultimatum.

Educate—or be extinguished.

The Institutional Betrayal

Academia has failed us. Universities crank out graduates who can recite OSI layers but can't secure a Linux box. Bootcamps promise six-figure jobs without ever teaching packet analysis or basic opsec hygiene. The result? A field full of checkbox champions and PowerPoint warriors who can pass audits but can't defend a thing.

We were promised security engineers. We got compliance clerks. We were promised leaders. We got influencers.

And the few who still fight in the trenches? They are outnumbered, outpaced, and drowning in ticket queues while the world burns.

The gap between theory and practice has become a canyon. And the bridge across it isn't curriculum—it's mentorship.

You Can't Outsource Literacy

Security isn't a SaaS subscription. You don't get it from a dashboard. You get it from repetition, failure, recovery. From nights spent chasing false positives and mornings spent parsing logs until your eyes bleed.

You learn by doing. By breaking. By building back stronger.

There is no shortcut. There is no substitute. There is no savior coming.

The Mandate for Mentorship

If you know more than someone else,
you are a mentor—whether you want to
be or not.

Teach others. Pull them up. Show
them how to wield 'tcpdump', how
to analyze 'strace', how to think
like an adversary. Not in webinars.
Not in corporate slide decks. In
command lines. In live fire. In the
uncomfortable mess of real systems and
real attacks.

Because if you don't teach them,
someone else will. And that someone
may not have their best interests—or
yours—in mind.

Be the one who answers the late-
night question. Be the one who reads
the CVE before the headlines do. Be
the firewall for the next generation.

Infosec Is a Trade, Not a Trend

We are craftsmen, not content
creators. We are practitioners, not

product evangelists.

Don't let TikTok experts rewrite the narrative. Don't let clickbait shape the curriculum. Don't let a generation be trained to memorize tools instead of understanding tradecraft.

Teach them how to threat model from nothing. Teach them how to defend without budget. Teach them to question everything.

Because in this world, the difference between survival and extinction is knowing not just how, but why.

The Price of Apathy

If we don't teach the next generation to care about privacy, they will grow up thinking surveillance is safety. If we don't show them how systems work, they will forever be owned by systems they can't control. If we don't arm them with knowledge, someone else will arm them with lies.

This isn't a plea for better
education policy. This is a
battlefield dispatch.

The war for digital autonomy will
not be won with budgets. It will be
won with minds.

So teach. Relentlessly. Not
because it's easy. But because the
alternative is unthinkable.

Educate.
Or extinguish.
Because if we do not pass the torch,
there will be no one left to hold the
line.

Chapter 9

When Bureaucracies Betray the Mission

> "Every system is perfectly designed to get the results it gets."
>
> — Paul Batalden

Bureaucracies do not fail by accident. They fail by design.

They are engineered to protect themselves, not the people they serve. They prioritize policy over principle, image over impact, delay over action.

And when they betray their mission—
when they forsake the vulnerable,
ignore the warnings, suppress the
truth—they don't apologize. They
obfuscate.

They don't patch. They pivot.
They don't fix. They reframe. Their
mission is not defense—it's optics.

Welcome to the enemy within.

The Illusion of Process

Process is supposed to create
accountability. But in the hands of
the apathetic, it becomes camouflage.
A way to hide inaction behind a wall
of procedure.

We've all seen it: The incident
report that leads nowhere. The
security ticket that gets closed
without resolution. The P1 that
becomes a backlog item. The policies
that exist solely to avoid liability,
not risk.

This is not incompetence. It is
indifference systematized.

They will hold three meetings to discuss what they could've prevented in one patch cycle. They will prioritize branding over breaches, tone over truth. They will spend more time perfecting their incident response memo than preventing the next incident.

They don't want security. They want the appearance of security—something they can PDF, email to a regulator, and forget by Friday.

The Institutional Immune Response

Try to fix something and watch the antibodies swarm.

Speak up and you're "disruptive." Demand real change and you're "not a team player." Raise alarms and suddenly you're the threat—not the breach, not the vulnerability, not the systemic rot.

They label you "abrasive" because they can't refute your facts. They say you lack "soft skills" because your questions make them uncomfortable. They rewrite narratives so that your warnings become insubordination, your diligence becomes defiance.

They don't want integrity. They want obedience.

And the moment you choose mission over politics, you are marked.

The Incentives of Failure

In most orgs, there's no real consequence for being wrong. Only for being early.

If you warn too soon, you're "paranoid." If you warn too often, you're "noise." If you warn correctly, but it makes leadership look bad—you're still wrong.

But miss the breach, fumble the response, cost millions in damage? Just say you followed the protocol. Blame the user. Blame the vendor. Blame the tool.

And keep your job.

Failure doesn't get punished. Truth does.

Security is a Threat to the Status Quo

Real security is inconvenient. It disrupts workflows. It challenges priorities. It requires action.

And bureaucracies hate action.

They want dashboards that look clean, not environments that are secure. They want compliance reports, not compromise assessments. They want "business continuity," even if that business is bleeding data from a thousand open wounds.

Security practitioners who care are labeled "zealots." Because nothing scares a stagnant system more than someone who is actually willing to trade their soul for a secure infrastructure.

What they call "overreacting" is often your correct reading of the threat. What they call "alarmist" is your refusal to wait for permission while data hemorrhages into the ether.

The Culture of Passive Sabotage

Sometimes the most dangerous threat actor is on payroll.

They're the ones who always delay your ticket. Who always need "one more meeting." Who always have a reason why it's "not the right time."

They sabotage by stagnation. They don't block you with denial. They drown you in delay.

This is weaponized bureaucracy. Slow enough to avoid blame. Fast enough to avoid accountability.

And death by a thousand deferrals is still death.

Refuse the Normalization of Neglect

It is not normal to: - Ignore repeated risk reports - Silence frontline analysts - Delay patching critical systems - Blame users while exempting leadership - Celebrate audits while breaches burn in silence

It is not normal to expect silence in the face of sabotage. It is not normal to spend more effort suppressing the whistleblower than stopping the exploit. It is not normal to lose sleep over liabilities that leadership won't even log.

This is not normal. This is betrayal.

And it must be named.

The Fight for Mission Integrity

Not all organizations are corrupt. But all are susceptible.

You must be the resistor. The one who documents everything. The one who speaks when silence is safer. The one who remembers that this job is not just about uptime—it's about people, systems, data, liberty.

Make yourself uncancellable by being undeniable. Build such technical clarity, moral authority,

and strategic precision that your
warnings can't be ignored.

Burn the receipts. Archive the
abuse. Chronicle the indifference.
Because one day, someone will ask what
happened—and your notes will be the
only honest record.

If they push you out for defending
the mission, let them know they lost
the only person who understood what
the mission was.

And when you go—go loud, go
documented, go right.

This Is the Line

We will not excuse incompetence. We
will not enable negligence. We will
not confuse process with progress.

We will call it what it is:
betrayal—quiet, sanctioned, and
systematic.

Bureaucracies may write the
policies.

But defenders write the history.

And history remembers those who

stood their ground.
   This is your ground. Stand it.

Chapter 10

The Way Forward

> "The best way to predict the
> future is to build it."

> — Alan Kay

We've surveyed the terrain. We've
named the threats, dissected the
myths, and reclaimed the ground that
should never have been surrendered.
Now we build.
This chapter isn't a conclusion—
it's a call to continuation. A
roadmap. A promise. A refusal to
settle for systems that surveil,

corporations that exploit, or bureaucracies that betray.

Make Sovereignty Scalable

You don't need to rebuild the internet. You need to reclaim your corner of it:

- Harden your machines

- Host your services

- Encrypt your data

- Teach your peers

One node at a time, we shift the balance.
Sovereignty scales through replication. Through mentorship. Through example.

Build Tribes, Not Empires

Empires seek control. Tribes seek cohesion.

Find your crew. The ones who get it.
Who document breaches in their sleep
and boot into rescue mode without
flinching. The ones who still care
when no one else seems to. Build with
them. Share code. Share threat intel.
Share failures.

Security is not solo work. It is
distributed resistance.

Adopt a Wartime Mindset

We are not in an era of peace. We
are in the age of ambient war—slow,
digital, unacknowledged.

Assume compromise. Assume
surveillance. Assume pressure.

But never assume defeat.

This mindset doesn't breed paranoia.
It breeds preparedness. It sharpens
your models, your tools, your resolve.

Make Defense Beautiful Again

Build systems that aren't just
hardened—they're elegant. Build

documentation that's readable. Build
UIs that empower.

Security shouldn't feel like
friction. It should feel like
freedom.

Legacy is Operational

You will not be remembered for what
you deployed. You will be remembered
for what you defended. For the alerts
you refused to ignore. For the red
lines you refused to cross. For the
people you protected when policy said
otherwise.

Leave behind playbooks, not
platitudes. Leave behind principles,
not policies.

Make your career a reference
architecture.

Own the Future or Be Owned by It

This is not a time for neutrality.
Not in the age of breached hospitals,
data brokers, and backdoored firmware.

If you don't architect a future worth inhabiting, someone else will build it for you—and it will not have your freedom in mind.

Freedom doesn't scale by accident. It scales because individuals take radical ownership of their stack, their systems, their ethics. It scales because someone refused to believe that vendor trust equaled safety, or that government silence equaled assurance.

You are that someone.

Security Is a Moral Imperative

This is bigger than red team or blue team. This is about what kind of world we want to live in.

Do we want a future where children's health records are ransom collateral? Where your refrigerator testifies against you in court? Where whistleblowers are hunted but surveillance profiteers get bonuses?

Then we build differently.

We harden not just code, but conscience. We patch not just binaries, but broken incentives. We refuse to tolerate negligence—at any layer.

Security, done right, is civil disobedience against digital decay.

Silicon Doesn't Have a Soul—You Do

AI doesn't have ethics. SaaS doesn't have loyalty. Cloud providers won't show up when the subpoena arrives. Your integrity must outlast the obsolescence of every tool you use.

The enemy is not just malware.

It's apathy. It's delay. It's every meeting where "low risk" gets rubber-stamped while red flags bleed in the logs.

The defenders of the future will not be the loudest voices in the room. They will be the quiet ones who refused to look away. Who kept the packet captures. Who knew when to walk—and when to go public.

This Manifesto Was Just the Bootloader

If you read this and think, "That was a strong read," then you missed the point.

This is a trigger. A primer. A blueprint for refusal.

Refusal to build systems you wouldn't trust with your child's data.

Refusal to play along with risk registers written in ink while breaches happen in blood.

Refusal to accept that "good enough" is ever good enough when lives, liberty, and truth are on the line.

And Now—Execute

Do not wait for permission.

Push the update. Deploy the fix. Encrypt the drive. Expose the breach. Teach the next defender.

The way forward isn't a strategy deck. It's a root shell. It's an audit trail. It's a generation of technologists who will not be bought,

broken, or silenced.
  History won't ask if you had access.
It will ask what you did with it.
  So choose your alias. Check your
configs. Update your threat model.
  And write the future in syslog.

The Work Begins Here

This is not the end. It's your
initialization script.
  Go build. Go defend. Go teach.
  And when the next breach hits—
because it will—you won't just
respond.
  You'll be ready.
  Because you chose root.
  And you built a future worth
defending.